WHEN THE EYE FORMS

WHEN THE

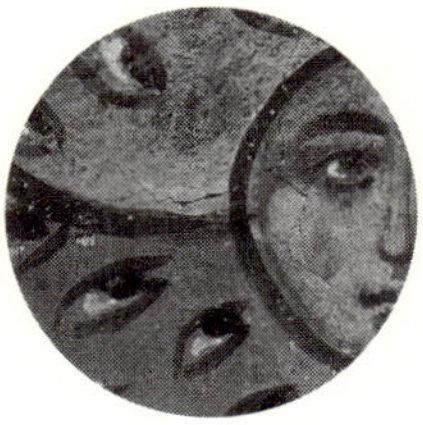

EYE FORMS

poems

{Dwaine Rieves}

Tupelo Press
Dorset, Vermont

Acknowledgments

A special appreciation to the following publications in which these poems are forthcoming or have been published:

Annals of Internal Medicine: "Holding On," "Edna in HerWords"
Baltimore Review: "Darkly, Lightly as Perspective"
Callaloo: "Stanton Hall"
Chelsea: "Sometimes They're Called the Connecting Tissues Diseases," "Posters"
The Comstock Review: "On the Sidewalk"
Crab Orchard Review: "Three Months After"
Crazyhorse: "Mississippi"
The Georgia Review: "Leaving"
JAMA: "Tracks"
The Journal of Medical Humanities: "Words"
Lancet: "Gay Mens VD Clinic"
Poet Lore: "Mardi Gras"
Potomac Review: "Double Design"
Rattle: "Jennifer"
River Styx: "Thrush, at 23; Aubade in E," "Athena This Year," "With Me, in Paradise"
Southern Poetry Review: "Out Burnt Gin Road", "Vaccine"
Sycamore Review: "Not the Same," "Well Baby"
Virginia Quarterly Review: "Shifts"
Vox: "At Night"
Wisconsin Review: "Another Song as The Man"

"Tracks" was reprinted in *Uncharted Lines*, JAMA, 1998.

For sharing places where we stayed until the poems came, I thank the nurses and patients of the Adams County Health Department in Natchez, Mississippi and the Whitman-Walker Clinic of Washington, DC. My gratitude, as well, to Margaret Donovan and Hillary Tham.

Copyright

ISBN-13: 978-1-932195-34-7
ISBN-10: 1-932195-34-3

Printed in Canada

First paperback edition October 2006
Library of Congress Control Number: 2006901779

Tupelo Press
PO Box 539, Dorset, Vermont 05251
802.366.8185 • Fax 802.362.1883
web www.tupelopress.org

TEXT DESIGN BY HOWARD KLEIN

COVER PAINTING: Seraphim with wings decorated with eye motif, fresco on exterior of the church, built in 1532 and painted in 1537. The Art Archive / Moldovita Monastery Romania/ Dagli Orti (A)

Contents

DUST WORLD

Places 1
Posters 2
World Book 4
Stanton Hall 5
To These Confederate Dead 7
In the Forest of Florid Remembrance 8
Floor 9
On the Sidewalk 10
With Me, in Paradise 12
At Night 14
Darkly, Lightly 15
Stone 16
Methodology 17
Someone 19
CPR 20

IN THE WOMB

Miss Welty in the Jitney on Fortification 22
Double Design 24
Tracks 25
Sometimes They're Called the Connecting Tissue Diseases 27
Gay Men's VD Clinic 29
Mardi Gras 30
Concrete Jesus 32
Well Baby 33
Here Waves 35
Shifts 37
Jennifer 38
Another Song 40
Mississippi 42
Curl 44

OUT WITH DOGS

Holding On 46
Leaving 47
Not the Same 49
Translations 50
Three Months After 51
Vaccine 53
Athena This Year 54
Graphic Before 56
Edna, in Her Words 58
Thrush, at 23 59
Out Burnt Gin Road 61
Assembly 63
Words? 64
Aubade in E 66

Notes 68

Locations and times—what is it in me that meets them all,
  whenever and wherever, and makes me at home?
Forms, colors, densities, odors—what is it in me that corresponds
  with them?

Walt Whitman, 1860, *number 22*

{DUST WORLD}

PLACES

working
the county health department
Natchez, Mississippi 1983
I was eager to leave—
career, cities, good places to live
One day
in clinic, a young mother—
I asked where you and your baby live
We *stay,* you answered, out Canal
And you meant it
Years later—answers,
yours came back in my Washington, DC
gay men's clinic
when one patient
you describing yourself—
your Mississippi drawl proud
of what a high end escort sees
Given the quick treatment
you stood to leave as the beeper,
yours sounded, as your face turned—
You, watching long distance numbers
I once saw in dreams
San Francisco, New York, Dallas
Doc, you wouldn't believe
the places I *stay*—
You beaming as
all my Mississippis forgave,
all my cities forgot
Places where we lived
I thought—
you telling yet how
we stayed You, I say

POSTERS

There must be a body inside—
black Lincolns have stopped
in the carriage lane and dressed up
men stand at the door. Across

the street, the funeral home is
Edwardian, its windows full body
size, black-lined and open. Inside,
it's probably viewing time.

But from my window, azaleas
dance about the veranda, April
unlayering red, like Lautrec's
madames, flipping saloon girl

skirts, working a syncopated twist
of upturned hips. *O How Pale She Is!*
Almost Paris. Outside,
Lady Avril fluttering among top hats

and sharp silhouettes, high kicking
a Grand Quadrille, her backup girls
in white underwear, guests wet
and poised with paper bills.

Lautrec must have known
the dead might want to be seen—
their hard tapping heels, each abrupt
decision tangible and done, told

in posters, five-colored and fired.
As commercials, the dead
beneath his hand, bodies in velvet
or naked behind two feathers,

their unstudied etudes danced
in the high silence of scenes,
names chalked on walls, faces
inside stores. This morning

azaleas dance out front while
parlors wait for everyone
to come inside, to stand before
the body, to see for themselves.

WORLD BOOK

We're in the new Bel Air, my grandmother
up front fanning, Mother driving,
me in the back with the black woman
who keeps our house. We're taking Tessie home

in a modern rocket, dirt clouds behind,
fields passing like World Book pictures,
ditches coughing up crickets while black-eyed
Susans laugh at the sky. These are new days.

Kennedy is alive and Ole Miss is not
in the news and people are suspicious but
summer is entering the wind, shirtless
and dirty, headed for Mexico. Still my grandmother
is all Mississippi and amazed.
Ain't those nigger navels something!

And there it hangs, summer's last question
big as my grandmother's face, there
as I poke her shoulder—my first thought that page
of an encyclopedia where I'd learned

those flowers were black-eyed Susans,
the proper name. Too much for the nudge,
she flushes—my grandmother red
as a planet she's never noticed. Under the U,
our wide universe it says, a two page spread

worlds beyond a Kodachrome but for the nights
creasing some bookish spine. Where lie
the acres I'll say Tessie stands within,
a woman at home in a field, face up
and neck craning, her jet black eyes
glad for rockets, dust and stars.
Yes Ma'am they're something,
she says and again,
Yes Ma'am they are.

STANTON HALL

Alone on the street
as Vicksburg fell,
the lady stood—her
white arms cold,

her porticoes smelling
of gunmetal and
spoiled cotton pods.
A door slammed,

its screen dissolved
and inside her, flies licked
horses, goats cleared
beds while hogs cuddled

in the parlor wedding cove.
Tell how she survived,
now on the tour, barefoot
and entered, pink in

her windows hanging
like twice chewed bubble
gum, girls hired
and smiling at the door.

Open the ivory shutters,
let her dead widows
crawl out. Call pilgrims
into peacock halls and

show the inside crust
of a propped up heart.
Inside, for five bucks
an hour, see it—her

slop jars dipped
in cologne, her nine-
foot doors still holding
their mahogany burl

while all Africa claws
at the fireplace walls, fingers
imprinting the bricks,
the fire barely used.

TO THESE CONFEDERATE DEAD

That birds would take over, someone
must have known, the soldier
(to believe the brass sign) centered
once on Main. Safer here, some committee
probably voted, the granite rebel
peacefully eroding. Who wouldn't love
to find the noble ignored? Just ask
the addicts and lovers meeting beneath
his feet. Or the birds—they might as well be
shitting on a post. Yes, birds could
cover an army. But they don't, generals
say as soldiers scatter like committees. Of course,
the generals have been here, medals
tossed among condoms and needles. At night, it all
glitters see, beneath the bushes.
                    Given our park's shot light,
the anonymous must be relieved
when the squawking begins. Tongued-plum
and toucan-blue—daybreak a committee might
never hear unless a granite gun fires. Such is
the dream, grit leeching free while hands
press even deeper the stones, years
caked over, fingers running where
the names come through.

IN THE FOREST OF FLORID REMEMBRANCE

Linger among the layers
of obedient colors,
fitful grins asking
                    what will it be—the futile
red, the surviving green,
a mutation's pink gadget?

Is it fear or guilt
that's always green?

        The collection stands livid,
leaves wrinkled like fathers,
stems clustered beneath their
sticky tongues.

Voices hide among those numbers.
Even so, you know their stories.
Their hands slap like seasons.

Engorged or dry, they reek
of expectations—faces
crumbling as winds get under,

petals listing before a phobic so
subtle no doubt
that first leaf never
swayed as the genetics touched.
                    Oh how surely
the green vein arced!

FLOOR

There were problems, of course,
the bent nail hit in angles until
the head came flush
with the board, surface
perfect. Finally, the usual colors.

Given quaint knots
and uncured grains, yellow and three-
quarters red, you might guess
boards ripped before a mule-powered
saw, the old animal
turning as wood slid
into hands, a lineman spitting before
the haul, each wagon tagged
with a tiny red flag.

One plank was lifted, then another,
two tacked together and so perfectly
aligned you want to believe
nothing went wrong, the animal
happy as it circled, ears lowered,
the man behind glad to be.
Like the carpenter

before a twisted nail,
an experienced hand was raised.
The rest is history, the way we know
it happened, planks so smooth
we're hollowed by beauty. Or so
we say when retelling our story, that time
of faithful animals, nails twisting
their heads. Polyurethaned

beyond a vintage
sheen, the floor asks to be touched—
splinters, burrs, hammers and belts,
one great blade spinning for days
before the turning took.

ON THE SIDEWALK

He's walking toward me.
Shoulders flared, back
tight, he mumbles and looks
about, the way a worker might
survey a loss, counting those
missing, telling those remaining
he is here now and in control.

Soft yellow, his suit wears an old
urine's haze, his necktie
the silk pucker of tropical flare.

Perhaps thorazine has sealed
his pores, has made his body
sweat-free and confident—
strutting through evening
heat, parading with a bishop's smirk.

Waving to neighbors, chatting
with trees, he smiles at strangers
and blesses the ones who stare.

Children wedge into mothers,
dogs resume their whines.

We pass, I nod, he smiles.
Maybe something was lobotomized
or some tough gland never
formed, the part that supports
a focus, a vision of body
as displayed self. At times,

I think contentment's no more
than misperception—a costume,
                hokey pockets.
If reality must be met
like this then let it

pass with the look of that man,
one who knows he's lost
an important part—a man
who meets the difference and walks
straight through the stranger
he sees.

WITH ME, IN PARADISE

The mystery of the world is in the visible, not the invisible.
Oscar Wilde

Yet surprise marks the sight of breath without
the body, light within an empty space, sweat
on the sheets and the bed thoroughly dry.
That sight's even promise, a rush
from the missing eyes of someone once
seen. As on the sidewalk this morning, three
men. One's walking a dog and the others
probably slept over, now headed home.
The first man's thumbing a retractable
leash—a button pushed and the dog's
pulled back. Cell phone dangling like a pistol
at the hip, the man's bull market, underwritten
with freshly edited smiles. Across the street,

the second man unlocks his car and drives
away, leaving like a priest on Prozac, the frown
predictably gone. But that last man's straight
from my catalog of once seen bodies. He's
HOT N DC, up from MAN 2 MAN, the escort
ads, his eyes locked on mine. Torso shot, a frontal
nude, he wears his web page pictures, a site
open still for hits. He's the Greek face
from a backside of paradise, his eyes
secure, uncensored, not curious or asking
why—a man walking alone in the early
morning hours, in those hours when
the women went to the tomb and left

in tears. There also, men may have passed
in silence, their faces occupied. Stories
were probably polished, parts removed, words
changed. Yet Gospels say on the hanging
night, Christ called for God then turned back
to the man at his side. And Scriptures note
morning's next mystery, stones rolled back,
a gardener, Mary taking one man for
another. Mistakes, delirium, the hurried
looks. At 16th and T, I remember all the bodies
I've never seen, those sudden eyes of a thief
entering paradise. Lover, open your hands.
On those cold nights, how can you
promise, how can you believe?

AT NIGHT

That sound is metal, a gate closing
at 2 AM, you upstairs in the window
watching shadows remove the garden.

It's all down there, willing to be
noticed or dreamed, what's being
taken or just now gone. The latch

fell quickly—the signal you assume
for shadows to move like lovers. Even
the streetlight's collusive, shadows

lifting their choices, root balls of
foxglove and coral birch, unsuspecting
spiders hanging veils above the empty

holes. The webs are good as draped about
your arms. So dark. Beside you,
a fly's lifeless shell, the black gonadal

eyes. From which you must say the night
was a sleepless delusion. Whatever
called you, only wind. This to

the garden's placid shadows and
missing lovers, to a web's writhing flies,
a light there glossing the wings.

DARKLY, LIGHTLY

Contained in the sky,
we are quiet, like Quakers,
enduring mountains

and serious towers,
our membranes ripped,

whispers leaving
our touched space.

Earth, keep your claws.
Up here whips make
lace, engravings fade.

Electric nets excite
our clouds,
steel digits tickle
our moons.

Our fragments
spin like rumors

you avoid.
No affluence, no
exotic cities. You

admire such medals,
our stars cored, our
pageants dependable as

the nights you try
to engineer. We dangle
like trinkets before

your breath. Breathe
in you say, blow
out and hear.

STONE

The strong ones refused scars
and welcomed hits, each muscle pulled
so far its springs could not return.
So their bodies turned to boroughs
where they—the strong ones—proceeded to die.

All the funerals featured
cannons and case-free bullets, no less than
two government tents. Church

bells wore leather muffs. Bound
and struck, half-tones left chimneys
clogged, store windows grease-lined
and noticed. Elegies knotted

the drapes, bishops telling how
strong ones made their places,
great cities and great speeches,
great highways where—

yes, a few lines were cut, a few
houses, the squat ones, razed.
Yet the product—

theirs—was a traffic that,
                    thank God, moved.

Whole nations adopted
their ways. So when it comes
to after-places, we can't blame senators
for begging to carry them through.

                    The strong ones.

We visit their stones to learn.
We leer before cracks
where great weights shifted
as the marble took,
our hands running the patina
while our faces wait.

METHODOLOGY

At the table where
skin's drawn and fascia
cleaved, the organs are

peeled. We're quiet
as the professor probes
methodically, our faces

following as he widens
the holes.
His hands move

quickly among questions
we're too astute
to suggest, his

fingers flicking enough
to fill a dead man's
chest and still he

looks, clipping scars,
yanking bands.
As the sternum is

threaded back into
position, we don't
ask that embarrassing

question about
the soul, how
certain necroses might

prove it exists as
conclusively as pans
of a clotted

growth—that is, until
it left for some
reason shortly

before the carcass
had to be hosed.
We say nothing as

the professor puts it
all together with
academic precision,

boards chalked
with unquestionable
reality—organs

we saw weighed,
a stainless scale,
systems named.

We saw the water
turned off, the chest
emptied. One insistent

stream. Always
dripping, Professor—
do they?

SOMEONE

Of luck, say what we didn't win.
We acknowledge our fixed
numbers and convoluted equations.

Say we accepted loss, our
lethal mutations, our pitiful
grins and ill-fitting wrinkles.

Say we kept at what kept us
here, bodies piling about
like a season's famous

leaves—incalculable numbers
someone must know. Say
we found the one leaf we

never wanted—veins red, its
heaven streaked as the summer
it became. O holy day.

CPR

Annie, Annie are you OK?
Of course, you're dead and we,
well at least we're certified to say so.

One Mississippi, two Mississippi...

Now do your lips taste
better post alcohol or bleach?
Or, given your plastic stare,
does sanitation even matter?

After all, if we're doing it right,
death has little to do with cleanliness
or any thing that living
demands, say sex or money.

And you, dead teacher, will
invariably agree, your cold lips erotically
spread as if you can take us always—

even if Mississippi dirty. You certifying
us all—our elbows straight, your
heart compressed and, if we're doing
it right, someone above us
always counting how long
another Mississippi takes.

{IN THE WOMB}

MISS WELTY IN THE JITNEY ON FORTIFICATION

Thought of leisure in decision, you one
cart ahead, weaving aisle cross aisle,
before corn flakes, then rice, lingering suggesting
there's rumor in labels, secrets on lids.
You wear that
matronly look of living alone, off-white
sweater above off-white blouse, your hair
a few wisps missing, the most part
womanly, your curls there. Still
          you pick and shake

the box and, for a moment
I hear Quakers whisper,
"oatmeal or sin?" Take Aunt Jemima's
turban. The red knot
          worn by widows or the rag
          tied roadside to a buggy's
shot wheel. Any story reconstructing
a banker's lie. Of experience,

say. Think woman blocking aisle,
her cart catty-cornered and most too polite
to push it aside.
          Or interrupt old ladies,
          or prod a woman.
          Remember capacity
or tension, the objectionable held
like you, a lady whose feet pan
          beyond her sandals,
heels on Galloway, toes bluff-off,
          there, as if spot-flashed
before a ten point buck.
Resistance, therapists say,
explaining important words, like speed.
To which their point: a pause
follows notorious names.
So Miss Welty,

if perception defends self
from self, our line's back half waits
while you flirt with a label's grin.
As though options know who must watch.
You, world hung from bone,
a back folded.
Your face immobile, the good
eye looking.

DOUBLE DESIGN

In 1949, the county supervisors concluded symmetry
should set the plan—the building new,

a clinic of identical, but opposite sides, each hall
with two water fountains, two commodes.

Left side, "White Only," right, no sign. Soon, two
wings reached out, walls high and gray,

their mixed pigments fixed in one agreeable
tone. In each hall door, etched glass

panes angled light, people seen as platoons moving
about, as shadows and smears. Cast

in cement, our building was built to endure, a pea
gravel shell, the foundation hard

and waiting. Years from now, when these wings
are just another fossil, chained off and on

display, can the concrete completely tell, the design
explain? No wonder certain statues

cry real tears, accidents divine our signs and
plagues cite opinion. When the world's

seen as a perfect palindrome, how else can the difficult
answer follow, but in language

of flipped inflections, whispers twisted into
tornadoes, unroofed tin wrapped

about the trees, clothes in a pond, all steeples gone?
This was our clinic's design.

The building is still wet with the thick scent of a storm
and cars being burned.

TRACKS

Once weighed for WIC, for program rules, mostly
black, mostly skinny, mostly children

line the clinic halls. The children can't see
rails spreading beneath their feet, those

steel connections feeding outer Dubuque
to inner Cleveland, all Enid to Orlando.

Neither can they see beyond these
walls—to Oklahoma, summer sand burying

the plank house, its family shrunken, alone.
And, over there, Montana, potbellied and quick,

a chill rattling among the pines, leaving them
bitten, done. Behind them isn't that

Dallas, the wire line dipping, dungarees strung?
There, it must be school's first day, because

down the block, bells buzz and charged voices rush
before a loud static slows as though someone hears

some shots and talk, they say, is of a showdown.
And in Detroit, the Little Engine That Could replugs

its cheeks, tagged to empty before Phoenix,
there to reload leftover cheese, too many stops,

too little time to gauge the gullies or estimate the odds.
All about this tangled land engines rumble, dust

piling, bridges listing, and chickens turning thin.
And here the line crackles with tiny fires, kids

popping sparks, their green wood barely burning.
Count it all, the cored cotton bud, the combine

and chain saw, the pilfered engine parts—with maids
from proud tobacco halls sweep the trashed receipts

from the courthouse steps, grab what's there
and paper the walls with all that's lost. Run

the ledger's tall columns until it's clear—the sum is
nothing and all that's left is unmeasurable heat.

SOMETIMES THEY'RE CALLED THE CONNECTING TISSUE DISEASES

Once naked and seen
from all sides, she is not
another skinny girl, another
family planning pelvic. Tall face, faintly
twisted spine, before me
a treble clef, left shoulder lifting,
right down and winding. Listen.

Her heart dances, a loose left heel,
its crystal buckle clicking. Run
her bones' supple harmonies:
an elbow's give, the carpals'
wheel. Think textbook etchings,
old French orchestras sounding
a slow unease, in the womb
an alloy's instrument replaced

with plastic, timbre mellowing
as it begins, that pull too pliable, too forgiving.

And your family, has anyone died?

One brother, a Kroger bag boy,
a parking lot, suddenly,
and everyone told like that, as when
it happens, an unzip
of lightning, a bridge washed
out, how operas end.
Unfile the coroner's conclusion—
a brother's aorta, one vessel open,
its walls splayed
beside a heart's abandoned rumble.
A tympany

maybe and no one warned.
Thunder-broken, can a brother's
peculiars predict what his sister may
share? Now

recheck the girl—her odd face, her
long chest, satin organs dangling
inside, a libretto's
accuracy captured,
each long note held
and chained to a soft leg. And one

day, when we learn
the formula for the moon's slow hum,
will the sound remain
as lovely, silver adagios
dusting a black-suited sky, one
unpredicted star dropping its assigned position—
arms softly falling
into night's perfect places?

GAY MEN'S VD CLINIC

And when he dropped his pants
I found there was no bedroom
                    or even a bed.
Maybe this is why some of the nits,
          the worried ones, have gone mad,
                              aghast at this nature—
a love of mobile homes,
          tiny trailers that weave and mat
                              as earth moves,
canyons rip
          and plains form.

Tentacles of pubic hair
spring and weave in a heaven raining dry
          pinpoint piles of mother blood,
          tumbled dust and ivory ribbed cocoons.
Shells and dander fall,
                    cobbling his underwear with rusty tiles,
          rolling with precision into cracks,
                                        along the seams.

All this, as if the social order in a henhouse,
          the gentility in the community urinal
and the decorum of a shower stall
          fuel gossip among a cult of tiny crabs,
          those sleepless members who must somersault
                              at the thought of moving on,
must spit their eggs
          among the high wires, frantically
                              crusting the sky
with brittle satellites.

In his sagging shorts, flaked
          scabs line the wrinkles
                    as an anxious galaxy swarms
above, its pace so frantic
                    each louse must
                    know the slight space
                    as home, the duration as joy.

MARDI GRAS

For the man who tested positive
just as the virus was named

At 16, I saw masks, a tight march
          of mercurial men, exotic wives,
          the queen's name taped to a Pontiac.
The parade was Endymion's—his well built boys,
          their fabulous confessions. Later

I learned how curbs were cleaned,
royalty exposed, the doubloons burned.
          In college I once dreamed
          of a god retaking his throne.
Bands were out
and all the delicate people followed.
First, one float with stout men

in a bull's belly; they pulled
and the beast bowed. Then, majorettes
          with rifles, contingents
of bare-breasted women and jungle boys.
          I liked the boys. Preachers shouted.
Crowds whispered. The Zulus went home.
I awakened but the boys continued,
          the thought a wayward krewe,

its muscled gods atop
          a Grecian barge, carnations tacked
to the wagon, the wagon hitched to my back.
I tugged
          and the barge-boys followed.
Streets narrowed and the float locked a curb.
I was 35, caught
          on a corner, alone with beautiful
monsters, an argument
with unambiguous sides. I reached
for vaults, the hiding caves,
the sound of my

unmarked name. I tell this story
and my patient smiles. A singe
        of needles, western blots,
        the anonymous test.
Sir, we've reached that point.
The sides are close, the ceiling heavy

and about us, these silent places.
        Pull when I pull, run
when I run.
Perhaps sparks and smoke—
flambeaux may fall. Inhale
armies and allies, their failed
        sulfates, the misted steel.
Let us enter
on the reflected sight: Endymion
        and his made-up men,
unmasked and following, one at a time.

CONCRETE JESUS

They're stacked five
to a row, outside
Bessemer, Alabama,
each Jesus to be
hauled away or stuck
where, like this, church-
yard bumpers scratch
his robes. Come Unto Me—
the banner says, letters
up there flapping
like bargains, none
saying—see, another
carpet store's gone bad.
Nor, what the believers
who took it over might
say—nothing is worth
more than a God poured
into body. No,
no miracle compares
even if cracks healed
in every Jesus between
here and Bessemer,
road-side stands closing
before the mark-downs were
changed. Inside, this body
or that, flaws so subtle
only God knows where
the concrete buckled.
Or why they left
this man to be
chipped and stained.
Hands open,
palms-up, the way
he looks away.

WELL BABY

On most visits you cuddled
a spot on the chart—
another check-up's milestone,

your age on the short axis, what
you could do on the other
and at the grafted cross

point, you. Systematically,
the chart tried to pin you
down—you crawling

over my shoulder, under
my ear—looking
for butterflies and chickens,

a lost shoe? Sometimes
I thought you saw me as easy
mystery, my body a clue

to this world's workings,
to whispers within
a skin's loose folds.

You fingered, gnawed
and tasted my face. I've
kept those records—you

on the rising line, mathematically
climbing out of your first
skin, its creek water

smell and uncertain tones of red.
Of course, knowledge is always
a two-faced bounty, a Janus

revealed, one face delight,
the other despair. The charts
say you climbed. Whereby

all mysteries begin:
here is the body, look
closely, see what it hides.

HERE WAVES

In this bed, waves
break and tug—
waves grabbing
metal-tossed light, the light
of four aluminum
rails. This bed
Mr. Phelps slobbers in—home
after a stroke, me bending
over. His words

stumble and fall.
His syllables detail the sky.
Off-lipped, he mouths
them, on tendons and bones, in air
sounds standing alone before
the sinew gives.
It's a turbulent discussion,

the debate maybe, of
cane or walker, an argument
for standing, for running

away. Below him,
waves are boundless,
tumbling still, they
refuse the sea,
silence cove
and sand, the stalking
calls. Here

Mr. Phelps takes his
hand to mine and tugs, pulling
my face before his.
The sand is wet.
It grips and even cracked,

the sky warns. As to horizons
totally new, Mr. Phelps
drags a half-anchored lip,

hearing it seems, what
I hear, saying *...no?...get away?*
I, the resolving sea. I,
the tin-struck sky.

SHIFTS

By 3 AM our unit's beyond coming or going, no
fluorescence, only bedside lamps and a voiceless
blue. It's an ICU sky, techs returning like comets,
you with your pressure falling into a range where
stars form so quickly we can't keep up with
the numbers.

Your nurse bandaged wet spots where skin
wouldn't hold. Your planets wept like blisters,
your heaven so swollen it had to give.
And as we sopped, you soaked through the world
we'd known, no family holding you down,
no orbiting lover.

We knew someone was talking
below, of what a body shouldn't endure, how it might
linger in what has to be pain, that system.
But you couldn't hear them, could you?
Not as your ankles turned to air, not as we dialed
the only number you left us, your voice answering
like so many dead keep doing these nights, your machine
implying you'd heard us all along.
Or so we took it

when you said who you were, your name and not
your number. You simply thanked us for the call.
Nothing, no requests, just that empty heaven
we refused to fill, your machine running
as the sky rose, your bandages pulled, the heavens
quiet as our hands passed over.

JENNIFER

In your panties, a small understain—
now unfolded and stiff,
you, five and on the table,
this cotton swab unlocking
summer, a valley's red breath—
touching where I don't
belong, inside, collecting fingerprints
from the face of cliffs.

In this world
every color is silent, touch
has no sound. Look,
purling over stone, beyond
these curves, a connecting
river, below.

Maybe such scenes are what
drove the old Monet
to paint his last stories in a deliberate
soft smudge. The uncertain edge rubbed
from each exact fact. Here

too, are facts—your father
with gonorrhea also,
with testimonials
confirming
where he was. Little Girl,

not etched in legal
charcoal and bone, today
we are struck in deep melting
shades, your mother's greased,
shock-black hair, your father's
big hands, his fingers coiling
in rouge and taupe, while outside,
a sheriff, blue.

Lie still, Little Girl,
the painting always ends.

Strokes detail only
time, this

world as Monet
finally saw his—
mounds of blazing
poppies sprouting
on the river's rough side,
their wide
faces fixed in the sun,
trembling.

ANOTHER SONG

He sits in the waiting room, his hands coupled
like a toy rattle—inside, coins are clinking.
Turning nine at twenty, he wears a syndrome's

face, his eyes two mysterious Mongolian
smiles, his heart a clamoring reed, its master
valve leaking. Suddenly, his coins fall—

nickels on the floor, rolling the length
of a man into the black space at the base
of a Coke box hum. A lurch, a whole body

drop, he follows, cheek slick with tile,
eyes below the old red machine, coning
a dark spot where his nickels stopped.

The waiting crowd mostly smiles, as if
to say they're all like this: chasing
little things, suspecting coins crawl,

shoes talk, a world flat, how they stare
into puddles, missing clouds, things reflected,
but watching like he watches, baseboard

places where dropped coins hide.
And down there, in a flea's lacy decay,
the dust world is probably frantic, anxiously

muffling each shiny coin with asbestos
and dried spider legs. I suppose
every world must redress its members,

must soften the inconsistent voice. Still,
no unsingable lyric survives in
a room of well-tuned ears, a room

where buses pass on the street outside,
their side-posters shouting perfection,
those smooth muscled boys sulking

in New York underwear. In this room,
so many are waiting—so many mommas
alive and watching. A man has spread

himself across the floor, arms reaching,
hands open. And in the underworld I know
there is a voice, a voice definitely singing.

MISSISSIPPI

For beauty is nothing but the beginning of terror,
which we still are just able to endure.
Rainer Maria Rilke

You were the good doctor.
You listened to problems,
put a stethoscope on the chests.
So frequently, by late day
you thought only
of the river, a bluff where
sky is washed
away. You stand above it
now, the river's broad lip
and on the other side, Louisiana
flat as an open hand.
You'd heard
the hearts all day. Eyes closed,
an empty river, a swollen stream.

Every squeeze came
bundled below, a chambered
fist, valves locked
so blood could only flow.
The waves raged
before your ears. You knew
what they were
because they kicked up
and peaked on the ribs,
limited by rhythm,
at least it seemed until
you listened so closely
you realized you'd fallen in.
All day you floated,
a low tone shimmering among catfish
bellies, the carcasses of goats.
You saw soggy, irregular
lilies, heard contraltos
in the upper range. The world

quivered like sisters.
You've heard too many problems.
They should not return.
Not to the bluff
of the river's beautiful
muscle, not to the heart
pulling you in.
The sky's incapable.
And so the river takes it.
You close your eyes
and hear the unending
descent, wind
coursing over heartless
river, the sky
that follows.

CURL

You find it swaying in
the morning—honeysuckle coiled over
the door's upper casings, the vine

littered with buds.
Such frayed but blossoming hair.

You can almost feel the turning—
in a single night, the weed-like coil
up from the alley's spilled
liquor, the urine
and ant-filled wrappers.
                    All night

the vine kept reaching
back then rising again, the next
a bit straighter, farther,

the whole endeavor continuing
despite the stem's ever famous
buds, despite the has-beens falling
beneath their weighty fame.

Really it's no surprise—

the famous confess while tumbling
from walls, the night's mortar
softening as tiny roots twist

inside. Tendrils up there groping
like hands in any dark alley, thoughtless
and raw. Most hidden,
those holding.

{OUT WITH DOGS}

HOLDING ON

When the need for survival becomes
an epidemic even the most alive feel sick.
Journal, 1984

In those days
centimeters matter.
Even calipers.
Every point will be
sharpened while politicians debate
precision, lifting
totals like hungry animals
sifting through sidewalk trash.
Yet in every

town—on every street,
an egg is hatched.

The new eyes blink,
then widely spread—
all engaged, all doubting—

LEAVING

First the father, then the mother,
taking turns by the month, bringing
the baby. A judge set the plan.
And after clinic, both parents do the same thing—

each holds the baby and sits
on the curb, then stands, and sits
again, stopping cars, thumbing a ride.
One day, the mother—wind opening

her hair, the baby snug at a breast
and peeping over her shoulder.
The mother tosses her head,
almost acknowledging something

a breeze offered, its whispers
of someone on the run, creeks being
dragged, men out with dogs.
The mother shifts her hips as if

she knows who got away. The baby
licks the mother's chest and rarely cries.
And a month later, the father.
See how his face turns

in the sun's under-burn, the baby
untangling in his arms. For some time,
the father stares to the opposite
curb, to a crushed squirrel,

its body stretched over asphalt
and drying, each claw still in
a frantic curl, as though grasping
something that can't be seen.

The baby is quiet and hard
to see, a blanket twisted, a face coming
out. Each time they leave, I picture
the three of them twenty years ahead:

the father and mother on separate streets,
still stopping cars; the baby
a woman who drives, a woman
who can't be found. And in

the scene, a shot-muffler sound,
the sound of someone floor-boarding
out County Line Road, the sound
of a car not stopping for anyone.

NOT THE SAME

What you heard,
someone knows

What you thought,
someone saw

Uncaught, a ball
rolls into traffic

You are the objection,
the taxi's jolted wheels

Someone screams

Yourself, your "me"
in the street,
running

still running

Sweet long notes,
beside you, branching tree

TRANSLATIONS

All morning I've listened
to sentences that sound
wrong, a contorted tense,
plural and singular confused,
sentences born the way
a woman's water breaks,
shoulders lifted,
arms shaking. Take Sylvia.
She's pregnant.

Today she told me
her father passed.
In Sylvia's chart, I said
her father died—words
accurate as the heart beat
of an unborn baby.
But Sylvia said her father
passed, not passed on
or passed away. Just *passed*—

as though some canyon had been
avoided, an escape made.

Passed.

She said it
calmly, as if
we were awakening
in the same dream.
A dream where she climbs
from bed and begins
to stretch—
legs tight, arms in the air,
every muscle pulling
both of us back
to the morning, back
to the world
we have to say.

THREE MONTHS AFTER

Strung about his belly, the scar's mathematics
score their Congo red, suture sockets
empty, cat gut dissolved. On this visit

he's fifteen pounds the record says,
a boy missing his stomach's outflow,
the viscera having twisted blue at birth.
Three months post-op, his mother now
eyes the scar. The question
is milk, the formula a special soy
she can't afford. She talks

of case-workers and 20 dollar cans,
how the county covers 26 of 30 days.
I look up and find the boy
limp within our numbers, his
mother lifting his face to hers.

"Little man, you're something,"
she says, pressing her nose into
desiccated skin—the boy's frog bone
legs and two-eyed worlds wandering
one to another. His mother

cups him on her chest as we walk through
a solution to the unpaid four—
                        a performance, an exhibit,
                        a story we'll take
to the Episcopal Women's Club, our
presentation following cards perhaps,
a talk on moist skin.

They're the ladies who do such
things—smiling women who pool
their special collections for the best
unfunded ways. And maybe
they'll take him, the boy their next
                        PROJECT 1984. I say

the ladies will be kind,
a brief visit, picture and plaque.

Still the mother's a question
so dry it's easily blown
away. Tell me, I want to say,
what is survival if not a spectacle of need?
            A camera, gardenias,
            the presidential hug.

Four extra cans and she wouldn't count
the forms she fingers before asking,
            "Little Man, ain't you something?"

And the boy dangles skinny
within her hands, the two another crowd-pleasing
picture, his momma's smooth, dark
hands holding him the way she knows
he'll need to be.

For the ladies, Dear.
Higher, please.

VACCINE

See, it's Margaret before
she's the big room nurse, the one
counting polio drops, months
between a DPT. Not alcohol
or CCs of expelled air but
honeysuckle at the doorway, someone
else or something you might guess Margaret
must be before collecting forms
or rechecking names. That's also
Margaret honking her rubber
dog so everyone knows she's
officially begun. *That's you,*
Sweetie. Have a seat.

And that's Margaret spreading
her knees, a baby girl locked
between while first momma
clamps deep both arms.
Honey, please. And that's
Margaret hiding her needle,
the girl's two-fingered flesh pulled up.
Just a little sting! And seconds
before skin's entered, it's
Margaret grabbing a breath
and holding it, holding it

until vaccine's in and needle's
out, until, gasping yet unfull,
the baby holds its scream the way
God surely hesitated before
Eve dropped to her knees asking,
just who do you believe?

Where went the air?
Who sent the bee?

ATHENA THIS YEAR

My neighbor, Annette, has gone
Greek. She's redone the garden
her husband planted last year.
He left, so she's uprooted hedge

and trashed the ivy. She's made
the garden a Grecian theme, white
on blue, an Aegean look of sun
and rock, her townhouse square

the small exception to the block's
summer pose, the faithful hosta
beds. From the sidewalk—
a rough mosaic, Annette's

central spread of crushed
marble, white facets turned
sharp points out, all cores
open. The rocks look hot,

at the end of day, their sides slick
as the gossamer from a bad
dream, a cantilevered glimpse
of some god or goddess

with diamonds for eyes.
Stone beside stone, each
stone barely touches another,
as if marble can hold no more

of monument or poise. Offered
is the crushed display—a statue's
breast or penis, the kore and kouros
dismembered. The outcome's

how rocks always fall, fault lines
open, stringencies like bones
marking where the bodies fell.
And at the edge of the yard—lavender

banks and blue-eyed Marys.
Every flower shares
the blue breath from
a fable's missing goddess,

her temple entered, her body
removed. She lives now
as the public show, decay
about the roots, each leaf

the simple acknowledgment
of what she's seen, how
she remembers the story.
Annette's chosen a garden

where myth endures—flowers
cresting the shards that fell
as a goddess watched, her
marble white, her flowers blue.

GRAPHIC BEFORE

"Charity Hospital Burns"
Natchez Democrat 1984

Charity— as you wanted
to keep her, brick and four-storied,
a wedding cake world stripped
of the raspy voice you imaged
she once magically assumed—Last Resort,

her name positioned among voyeuristic
desires, those noble voices drawing you
back to the door where barred
windows were the only cryptics
you craved to explore. You knew how
public suns go down,

but none as perfectly alone
as that opening like a discarded lover
over Charity's dank halls. That evening
you stepped up and found Charity's cedar
    heart running, her rooms
    swept out, her plaster ignored.
    You watched and planted

ghosts inside, what you had to see—
white girls loading the gurneys,
dangerous men, old blood spotting her floors.

Yours was a world of gunshots,
babies and broken jaws—
a Charity of blessed odors
and painless rubbing, a gold-toothed nurse calling
    Baby, come see
    behind the building's beautifully locked door,
your hesitation the same
reluctance you share now

with all voyeurs, even those
politicians walking among Charity's burned-out
walls, a full color photo
capturing what every spectator fails
at best. You and those men forever
kicking at the ashes, at the world
you had to leave—
steel clanking, ash settling,
a gold-toothed nurse calling
from every locked door.

EDNA, IN HER WORDS

On the file's front page, she's assigned
no diagnosis, no Latin offshoot
to name the disease on a line.
It's the reason we return to her pain,

unraveling symptoms on alternate months,
restating the signs, every cul-de-sac
and every algorithm's arrowed option ending
with an answer of nothing wrong, every

duplicate test and well-inked consultant
agreeing—her spine's clicking does not exist,
the corrugated rub's simply not there.
Within all pain: an upper spot

unattended, a warm under-wall
and an iron rail splitting the middle.
It's the picture Edna uses to describe
her daughter, a daughter, she says, turned

to metal. In Edna's words,
she sometimes meets her daughter
in Wal-Mart and the daughter never speaks,
she simply pushes her cart away. At first

tangential—but Edna says it as though
her ear's on the rail when a name's shot through,
like metal struck in an empty place,
like a diagnosis missing its line, her line.

THRUSH, AT 23

Today's last patient
comes with his mother,
with fever, a throat on fire, the look
of someone locked away,
smelling smoke.
We don't know
this is the year of answers—a virus.

His mother says
he came home, moved
back in. I'm learning how it works
when mothers bring
their adult kids—
and this evening, I probe
the young man's throat, his mother
over my shoulder, looking
down him.

"Open wider."

Maybe she sees it, the white coral
crust, cobbled and glistening,
as if its lumpy parts must
toss their silver before
collapsing, sexless
atop pearl beds. And caught
by the light,

the patch of thrush is wet
and full, like the fungus bodies
were found, parked and hugging
each other, mid-swoon
on a summer night.

            Maybe she hears it,
the tactile sounds, suspended
within
his jaws—a rustle, the slow collecting
            of thrush; its plaster ghosts
mumbling, stunned.
She leans
as if she must go to the tip of the light

and see for herself what distance
contains—
            crystal hills, unmade beds,
voices and names.

Then she stops and backs away
            and he closes
his mouth and they are silent
and fixed—in
            a new place
and waiting for me.

OUT BURNT GIN ROAD

To your name on the note,
the nurse added
pregnant—car broke down.

I called, promised
to be at your place,
by 10. Out Burnt Gin Road,
beyond cut bodock and black tupelos,
I find you sitting
on a front yard stump. Here too
there's mist, your face
penumbral, your eyes
telling, like the hunted, how
it feels to be found.
I've seen your face

before, a high Renaissance
painting, a stallion breaking
across a bog,
on its back a plump woman,
her grip hidden
in the uncropped mane.
That scene probably some painter's

escaping Diana,
a woman the fog can't keep, her
eyes on the water, her face one
beginning of submission.
Reared back

in the picture, the horse's head,
its eyes locking the opposite
shore. That,
though the horse must feel
the woman's hold
softening, the coming ridge
almost touched.

Out Burnt Gin Road,
among fresh cut trees
and jacked-up cars,
I find you watching.
You in an early showing
time, in that trimester
when the eye forms.

ASSEMBLY

The brain, I said, explaining
the largest sex organ to teenagers
in a darkened hall. I'd come to
illustrate disease. Red flesh,
white cheese, sores bigger
than a book. I clicked and
watched the eyes. You name
it—the heartbreak of VD, warts
on 3D slides, wilted condoms
emphasizing God knows
what. Tall windows and tall
faces, even I felt taken over, my
brain peeled like a massive banana.
Was the concept virtue or pain?
Or simply sunsets and blobs,
two great blisters smiling?
Postcards almost, mangoes
or melons, whole rows
sun-up and ripe, hairy tails
begging to be pulled. Close-up
photos fell from a sun so close
I wanted to believe I held its
beam—another slide rising as it did
above the heads, a horizon quiet
but for the hum of teenagers,
row after row, all plump
and sun-dazed, every
melon turning in its field.

WORDS?

At night, in my bed, a rip-shirted
world still breathes, its muscle
    matched to mine, its rhythm

in my back, one slow heart
unschooled and warm. Come
    inside, check your coat,

remove your shoes—or keep them,
it is so risky, the intonations, the accents,
    the manipulations of day, its

science and short sun, its voluptuous,
sudden moon. A moon so focused
    I recorded its holes, the few

mumblings it could give. I saw these
things and kept their answers.
    Study the scenes—

on-call, a night's receiving room,
the dead woman, her chest open,
    my hand inside. That was

the night I took back those
early sayings, their recommended lists
    and ready places.

That night I crawled down
from the stacks. I kicked Dr. Osler
    and opened his tie, finding

old Halsted, scalpel raised and in
my face. And later, with Blalock—
    we chewed opium and talked

about pain. Remember
these nights, the short tempered
    streets. I sleep with this

world and speak its words,
the unstruck, the found, sentences
        pale and sweating, that

whole language of make-do
words, every syllable bare chested,
        beside me still and breathing.

AUBADE IN E

The wet voice is soft, a slow transfer of touch,
breath returned to the sun, now breast plated and tense
with spears. So much opening sex, hushed mandates
and decisions—out here the living can't sleep alone.
The proof's one communal array of baby
doves. Yet a voice says questions survive in the unseen,
in a seed's cracked shell, the red core uncontained.
What's left once any lover's gone? It's the soul lifting
both dirty hands, a fall back of flesh to salt,
jeans on the floor, a slow walk home. Today's the first
day of pink lips and unapologetic green,
spring's opera staged within every dying crocus
head. I hear a thousand awkward sounds, the sweet
silence of the body abandoned, a voice the body loved.

Notes

Posters (page 2)	*O How Pale She Is!* is a song from Lautrec's Paris.
Stanton Hall (page 5)	Stanton Hall, built in 1857, is one of the largest slave-built mansions in Natchez, Mississippi.
Miss Welty in the Jitney on Fortification (page 22)	Eudora Welty shopped at this grocery store in Jackson, Mississippi.
Tracks (page 25)	*WIC* is a nutrition program for women, infants and children.
Sometimes They're Called the Connecting Tissue Diseases (page 27)	Diseases such as Marfan's Syndrome, an inherited disorder of skin, blood vessels and joints are connecting tissue diseases. Antonin Marfan, a French pediatrician, first described Marfan's' Syndrome.
Mardi Gras (page 30)	*Zulus:* An African-American krewe in New Orleans Mardi Gras celebrations.
Another Song (page 40)	*Mongolian:* After "mongoloid."
Vaccine (page 53)	*DPT* is a childhood vaccine containing diphtheria, pertussis and tetanus toxoid.
Graphic Before (page 56)	Natchez Charity Hospital overlooked the Mississippi River and the Natchez Community Cemetery. Abandoned, it burned in 1984.
Words? (page 64)	*Osler, Halsted, Blalock:* were historically prominent physician-professors.